Bring Your Own Device (BYOD) to Work

Bring Your Own Device (BYOD) to Work
Trend Report

Bob Hayes
Kathleen Kotwica

ELSEVIER
AMSTERDAM • BOSTON • HEIDELBERG • LONDON
NEW YORK • OXFORD • PARIS • SAN DIEGO
SAN FRANCISCO • SINGAPORE • SYDNEY • TOKYO
Security Executive Council

Elsevier
The Boulevard, Langford Lane, Kidlington, Oxford, OX5 1GB, UK
225 Wyman Street, Waltham, MA 02451, USA

First published 2013

British Library Cataloguing in Publication Data
A catalogue record for this book is available from the British Library

Library of Congress Cataloging-in-Publication Data
A catalog record for this book is available from the Library of Congress

ISBN: 978-0-12-411592-7

This book has been manufactured using Print On Demand technology. Each copy is produced to order and is limited to black ink. The online version of this book will show color figures where appropriate.

Transferred to Digital Printing in 2013

CONTENTS

EXECUTIVE SUMMARY

Bring Your Own Device (BYOD) is the practice of employees bringing personally owned mobile devices (e.g., smartphones, tablets, and laptops) to their place of work, and using those devices to access company resources such as email, file servers, and databases. The general consensus is that BYOD is coming (or is already here in many instances), so it's inevitable that all companies will have to establish a BYOD program. BYOD will, and does, present unique issues to consider and address, but addressing these matters in light of existing practices and postures regarding data privacy, confidentiality, security, and acceptable use will facilitate proper program decisions and a successful BYOD implementation.

Building and implementing a successful BYOD program requires a cross-functional effort with contributions from information technology, human resources, finance, legal, security, and business operating teams; the effort should focus on identifying and defining program needs, requirements, and objectives that will address the unique circumstances of company business operations. The impact of BYOD on employee productivity and the pros and cons of instituting a BYOD program will be discussed. BYOD must-haves, challenges, and risks and liabilities are also explored. This report is a valuable resource to any security professional who is considering the implementation of a BYOD program.

WHAT IS A TREND REPORT?

A trend report is a document that highlights emerging and fast growing trends with significant impact for corporate security and risk management. Based on first reports from initial responders to the issue and confirmed by research, these reports help industry leaders and practitioners learn the key elements of an important topic, and provide insight, guidance, and options for applying what has been gleaned from a real-world environment. These reports can be utilized by mid- to upper-level security managers, instructors at institutions of higher education, or by human resources professionals in training sessions.

Research Findings

1.1 PRODUCTIVITY

Data from various recent surveys associated with BYOD and productivity in the workplace are listed below. Due to the relative newness of this issue, there is not a lot of research to be found outside of that from vendors.

- 81% of college students believe they should be able to choose the devices they need to do their job ("Cisco 2011 Annual Security Report," December 2011)
- 58% of IT decision makers think mobility and consumerization can deliver increased productivity and efficiency ("BT Assure: Rethink the Risk Summary," April 2012)
- 84% of IT decision makers believe companies allowing employees to use personal devices for work enjoy a competitive advantage ("BT Assure: Rethink the Risk Summary," April 2012)
- Nearly 4 in 10 of enterprise organizations surveyed have a history of BYOD-related breaches in security ("BT Assure: Rethink the Risk Summary," April 2012)
- 62% of surveyed enterprises pay for employee devices and voice data plans ("BYOD and Virtualization, Insights from the Cisco IBSG Horizons Study," May 2012)

- 72% of survey respondents are already formally supporting a BYOD program ("Good Technology State of BYOD Report," October 2011)

1.2 PROS AND CONS

1.2.1 Pros

- Improved user experience due to device familiarity and device singularity (one device)
- Potential hardware cost transfer from company to employee
- Improved work from anywhere/anytime opportunities
- Workplace draw for young professionals, e.g., "Best Places to Work"
- Increased workplace productivity

1.2.2 Cons

- Potential loss of company purchasing power related to a reduction in bundling of traditional technology services (software/hardware/usage)
- Increased difficulty for in-house IT user support due to multiple platforms and devices
- Hardware and software compatibility issues with device to organizational software and infrastructure
- Increased mix of personal and company information; blurring the lines of company vs personal property
- Introduction of new data security/privacy threat opportunities

1.2.3 Pro/Con

- Hardware refresh: more frequent device upgrades containing the latest features and capabilities (depends on the equipment life cycle of the specific company)

1.3 MUST-HAVES

A clearly communicated BYOD program should contain the following elements:

- Definition of program eligibility: people, devices, and data/program/application categories
- Who pays and how? Device and plan usage (employee or company, full or partial, stipend [%] or expense)

- Who provides IT support? In-house or employee/device carrier?
- User responsibilities (rights, privileges, expectations)
- Company rights and privileges
- Security requirements
- Employee user agreement to manage expectations, clarify responsibilities, and address potential legal, employment, and privacy-related issues
- A Mobile Device Management (MDM) program (onboarding, tracking, identification, management)
- Employee awareness training and certification program
- A mobile security audit program (for tracking devices, users, and applications)
- Ability to remotely find (GPS), wipe, and/or kill all lost, stolen, or terminated devices
- Clear device and operational security requirements

1.4 CHALLENGES

When instituting a new BYOD program, companies may experience challenges related to the following:

- Gaining physical access to devices
- Installing and managing device security software upgrades and patches
- Wiping devices of company data associated with user-initiated device upgrades
- Recovering proprietary information from terminated devices and third party (cloud) storage
- Confiscating devices associated with investigations and/or legal discovery/holds
- Enforcing and monitoring acceptable use, data storage, software and hardware security compliance

1.5 RISKS AND LIABILITIES

The following risks and liabilities are possible when instituting a new BYOD program:

- Loss of security control; greater risk related to intrusion control/detection and malware susceptibility
- Data breach risk and related legal liabilities

- Managing and maintaining legal, regulatory, and contractual obligations
- The increased risk of external software applications (apps) introducing malware
- The risk for exposure of company data due to increased device use for non-business-related activities

REFERENCES

BT Group plc. (2012, April). BT Assure: Rethink the Risk Summary. Retrieved from <http://globalservices.bt.com/static/assets/pdf/campaign/security_that_matters/BT_Assure_Rethink_the_Risk_summary.pdf>.

Cisco Systems, Inc. (2011, December). Cisco 2011 Annual Security Report. Retrieved from <http://www.cisco.com/en/US/prod/collateral/vpndevc/security_annual_report_2011.pdf>.

Cisco Systems, Inc. (2012, July 17). Cisco Bring Your Own Device: Device Freedom Without Compromising the IT Network. Retrieved from <http://www.cisco.com/en/US/docs/solutions/Enterprise/Borderless_Networks/Unified_Access/byodwp.pdf>.

Cisco Systems, Inc. (2012, May). BYOD and Virtualization: Top 10 Insights from Cisco IBSG Horizons Study. Retrieved from <http://www.cisco.com/web/about/ac79/docs/BYOD.pdf>.

Cisco Systems, Inc. (2012, May). BYOD and Virtualization: Insights from the Cisco IBSG Horizons Study. Retrieved from <http://www.cisco.com/web/about/ac79/docs/re/IBSG_Horizons_BYOD_KeyInsights.pdf>.

Good Technology. (2011, October). Good Technology State of BYOD Report. Retrieved from <http://media.www1.good.com/documents/Good_Data_BYOD_2011.pdf>.

Good Technology. (2012). Bring Your Own Devices Best Practices Guide: A Practical Guide for Implementing BYOD Programs at Your Organization. Retrieved from <http://www.welcometogood.com/byod3/_assets/pdf/byod_best_practices.pdf>.

Good Technology. (2012, January). Bring Your Own Device: Individual Liable User Policy Considerations. Retrieved from <http://www.welcometogood.com/byod/byod_policy_wp.pdf>.

Tangoe, Inc. (2011, January). ProfitLine Mobility Whitepaper: The Hidden Risks of a "Bring Your Own Device" (BYOD) Mobility Model. Retrieved from <http://i.zdnet.com/whitepapers/Profitline_The_Hidden_Risks_of_a_Bring_your_own_Device_BYOD_Mobility_Model_1_19_2011.pdf>.

ADDITIONAL RESOURCES

Aberdeen Group. (2011, July). Prepare Your WLAN for the BYOD Invasion. Retrieved from <http://research.aberdeen.com/internetcontent/somoclo/0161-7262-AI-WLAN-BYOD-AB-08.pdf>.

Aerohive Networks, Inc. (2012). The iEverything Enterprise: Understanding and Addressing IT's Dilemma in a Bring Your Own Device (BYOD) World. Available from <http://thejournal.com/whitepapers/2012/aerohive_12a/the-ieverything-enterprise/asset.aspx?tc=assetpg&returnkey=IX9I3Hhzi2ni0Fr7eQMXW2Bvnah10MzG&LCSKey=38807564-2f0e-4f7b-b053-725eaa360e62>.

Aruba Networks, Inc. (2011). Bring Your Own iPad to Work. Retrieved from <http://www.arubanetworks.com/pdf/technology/whitepapers/WP_Bring-Your-Own-iPad-to-Work.pdf>.

IBM Corporation. (2011, September). IBM X-Force Mid-year Trend and Risk Report (section II, page 68). Retrieved from <http://public.dhe.ibm.com/common/ssi/ecm/en/wgl03009usen/WGL03009USEN.PDF>.

IT Business Edge. (2011). BYOC Acceptable Use Policy Template. Available from <http://www.itbusinessedge.com/itdownloads/byoc-acceptable-use-policy-template/89199>.

Knowledge and Information Services, Office of the Chief Information Officer, Ministry of Citizens' Services, British Columbia. (2010, October 18). iPhone, iPad & Bring-Your-Own-Technology Practice in the Workplace. Retrieved from <http://www.cio.gov.bc.ca/local/cio/kis/pdfs/iphone_ipad_byot.pdf>.

Telesoft. (2011, July). Corporate Mobile Policy Template. Retrieved from <http://www.telesoft.com/sites/default/files/pdfs/whitepapers/Telesoft_whitepaper_mobilepolicy.pdf>.

The SANS Institute. (2008, May 29). Security Policy for the Use of Handheld Devices in Corporate Environments. Retrieved from <http://www.sans.org/reading_room/whitepapers/pda/security-policy-handheld-devices-corporate-environments_32823>.

The SANS Institute. (2009, June). Personal Digital Assistant Audit Checklist, Version 2.2. Retrieved from <http://www.sans.org/score/checklists/PDA_Audit_Checklist2.doc>.

ABOUT THE AUTHORS

Bob Hayes has more than 25 years of experience developing security programs and providing security services for corporations, including eight years as the CSO at Georgia Pacific and nine years as security operations manager at 3 M. His security experience spans the manufacturing, distribution, research and development, and consumer products industries as well as national critical infrastructure organizations. Additionally, he has more than 10 years of successful law enforcement and training experience in Florida and Michigan. Bob is a recognized innovator in the security field and was named as one of the 25 Most Influential People in the Security Industry by *Security Magazine*. He is a frequent speaker at key industry events. He is a leading expert on security issues and has been quoted by such major media outlets as the *Wall Street Journal* and *Forbes*. Bob is currently the managing director of the Security Executive Council.

Kathleen Kotwica has a PhD in experimental psychology from DePaul University and has had a career as a researcher and knowledge strategist. Her experience includes positions as information architecture consultant at a New England consulting firm, director of online research at CXO Media (IDG), and research associate at Children's Hospital in Boston. She has authored and edited security industry trade and business articles and has spoken at security-related conferences including CSO Perspectives, SecureWorld Expo, ASIS, and CSCMP. In her current role as EVP and chief knowledge strategist at the Security Executive Council she leads the development and production of Council tools, solutions, and publications. She additionally conducts industry research and analysis to improve security and risk management practices.

About Elsevier's Security Executive Council Risk Management Portfolio

Elsevier's Security Executive Council Risk Management Portfolio is the voice of the security leader. It equips executives, practitioners, and educators with research-based, proven information and practical solutions for successful security and risk management programs. This portfolio covers topics in the areas of risk mitigation and assessment, ideation and implementation, and professional development. It brings trusted operational research, risk management advice, tactics, and tools to business professionals. Previously available only to the Security Executive Council community, this content—covering corporate security, enterprise crisis management, global IT security, and more—provides real-world solutions and "how-to" applications. This portfolio enables business and security executives, security practitioners, and educators to implement new physical and digital risk management strategies and build successful security and risk management programs.

Elsevier's Security Executive Council Risk Management Portfolio is a key part of the **Elsevier Risk Management & Security Collection**. The collection provides a complete portfolio of titles for the business executive, practitioner, and educator by bringing together the best imprints in risk management, security leadership, digital forensics, IT security, physical security, homeland security, and emergency management: Syngress, which provides cutting-edge computer and information security material; Butterworth Heinemann, the premier security, risk management, homeland security, and disaster-preparedness publisher; and Anderson Publishing, a leader in criminal justice publishing for more than 40 years. These imprints, along with the addition of Security Executive Council content, bring the work of highly regarded authors into one prestigious, complete collection.

The Security Executive Council (www.securityexecutivecouncil.com) is a leading problem-solving research and services organization focused on helping businesses build value while improving their ability to effectively manage and mitigate risk. Drawing on the collective knowledge of a large community of successful security practitioners, experts, and strategic alliance partners, the Council develops strategy and insight and identifies proven practices that cannot be found anywhere else. Their research, services, and tools are focused on protecting people, brand, information, physical assets, and the bottom line.

Elsevier (www.elsevier.com) is an international multimedia publishing company that provides world-class information and innovative solutions tools. It is part of Reed Elsevier, a world-leading provider of professional information solutions in the science, medical, risk, legal, and business sectors.